The INSIDE GUIDE

STEM SUPERWOMEN

# Katherine Johnson

By Katie Kawa



Cavendish Square

Published in 2026 by Cavendish Square Publishing, LLC
2544 Clinton Street Buffalo, NY 14224



Website: cavendishsq.com



Library of Congress Cataloging-in-Publication Data

Names: Kawa, Katie, author.
Title: Katherine Johnson / Katie Kawa.
Description: Buffalo, New York : Cavendish Square Publishing, [2026] | Series: The inside guide : STEM superwomen | Includes index.
Identifiers: LCCN 2024056593 | ISBN 9781502674623 (library binding) | ISBN 9781502674616 (paperback) | ISBN 9781502674630 (ebook)
Subjects: LCSH: Johnson, Katherine G.–Juvenile literature. | African American women mathematicians–Biography–Juvenile literature. | Women mathematicians–United States–Biography–Juvenile literature. | United States. National Aeronautics and Space Administration–Officials and employees–Juvenile literature.
Classification: LCC QA29.J64 K39 2026 | DDC 510.92 [B]–dc23/eng/20250111
LC record available at https://lccn.loc.gov/2024056593

Editor: Katie Kawa
Designer: Deanna Lepovich

The photographs in this book are used by permission and through the courtesy of: Cover, pp. 12, 13, 15, 16, 29 (left) Courtesy of NASA; p. 4 Katherine Johnson 1983/Wikimedia Commons; p. 6 Mark Winfrey/Shutterstock.com; pp. 7, 28 (top right) Everett Collection/Shutterstock.com; p. 8 Chad Robertson Media/Shutterstock.com; p. 9 NACA seal (cropped)/Wikimedia Commons; pp. 10, 22 Bob Nye/NASA; p. 14 Airplane vortex edit/Wikimedia Commons; p. 18 Alan Shepard 1961/Wikimedia Commons; p. 19 Friendship 7 (big)/Wikimedia Commons; p. 20 Langley Research Center/NASA; p. 21 Mae Carol Jemison/Wikimedia Commons; p. 24 NG Images/Alamy Stock Photo; p. 25 Katherine Johnson at the Katherine G. Johnson Computational Research Facility/Wikimedia Commons; p. 26 Kim Shiflett/NASA; p. 27 NASA Image Collection/Alamy Stock Photo; p. 28 (top left) GPN-2000-001027/Wikimedia Commons; p. 28 (bottom left) Kathy Hutchins/Shutterstock.com; p. 28 (bottom right) Salma Bashir/Shutterstock.com; p. 29 (right) SeventyFour/Shutterstock.com.



CPSIA compliance information: Batch #CSCSQ26: For further information contact Cavendish Square Publishing LLC at 1-877-980-4450.

Printed in the United States of America

# CONTENTS

Katherine Johnson has inspired many women to turn their interest in STEM—science, technology, engineering, and math—into a career.

# AN INSPIRING STEM STORY

Chapter One

Sending people into space is considered one of the most impressive achievements in the history of STEM. However, the women who helped make spaceflight possible went unrecognized for many years.

Katherine Johnson was one of those women. Then, in 2016, a book and a movie called *Hidden Figures* were released. They told the story of the Black women whose knowledge of STEM sent Americans into space. This included Katherine's story, which has inspired many women and girls, especially Black women and girls, to follow her path to STEM success.

**Fast Fact**

The full title of the book that brought Katherine's story into the spotlight is *Hidden Figures: The American Dream and the Untold Story of the Black Women Who Helped Win the Space Race.* It was written by Margot Lee Shetterly.

## A Natural with Numbers

Katherine was born on August 26, 1918. She grew up in White Sulphur Springs, West Virginia. Her parents worked hard to build the best life possible for her and her three siblings. Katherine's father had many different kinds of jobs, and her mother worked as a teacher.

**Fast Fact**

Katherine's full name when she was born was Creola Katherine Coleman.

From an early age, Katherine was a natural with numbers. She once said, "I counted everything. I counted the steps to the road, the steps up to church, the number of dishes and silverware I washed ... anything that could be counted, I did."

## Skipping Grades

Katherine's interest in numbers made her a very successful math student. She also **excelled** in her other classes. She even skipped grades because she learned so quickly and knew so much.

**White Sulphur Springs is the location of a resort called The Greenbrier, shown here. Katherine's father worked there as a bellman, or someone who carries people's suitcases for them.**

# SEGREGATION IN SCHOOLS

Katherine grew up at a time when segregation was legal in many parts of the United States. Segregation was the forced separation of Black and white Americans. Because of segregation laws, Black Americans were forced to use separate water fountains, restaurants, bathrooms, and schools.

Katherine had to travel far to go to high school because there was no high school for Black students in White Sulphur Springs. Segregation in education was only supposed to be legal if the schools for white and Black students were equal, but this was clearly not the case.

Segregation in U.S. schools was declared unconstitutional, or against the U.S. Constitution, in 1954. However, it took many years for all the schools in the country to become integrated, or open to students of all races.

Katherine's parents wanted her to go to the best high school available, so she went to school in Institute, West Virginia, which was a long way from home. She started high school at the age of 10 and then went on to West Virginia State College. Katherine took every math class the college offered, and she finished college when she was just 18 years old.

**Fast Fact**

**One of Katherine's teachers created a new math class just for her. It was all about using math to understand space!**

## A Groundbreaking Graduate Student

After Katherine finished college in 1937, she started working as a math teacher in Virginia. At the time, teaching was one of the few jobs available for Black women who loved math.

**In 1939, West Virginia University (*shown here*) integrated its graduate school, which students can attend after graduating from a four-year college. Three Black students began their graduate studies there that year: two men and Katherine.**

Two years later, Katherine went back to school. This time, she studied math at West Virginia University in Morgantown, West Virginia. Katherine was the first Black woman to take graduate, or advanced, classes at that school.

## A New Opportunity

Katherine left school to start a family. The same year she started graduate school, she married James Goble. Motherhood came next for Katherine, and she and James had three daughters.

Later, Katherine went back to teaching. However, she really wanted to be a **research mathematician**. In 1952, a relative told her about the National Advisory Committee for Aeronautics (NACA), which was hiring Black women to work as mathematicians. Katherine got a job there the next year and never looked back.

People who worked for NACA, like Katherine, dealt with aeronautics, or the science of flight.

The late 1950s and 1960s were an important time for space exploration, and Katherine played a major part in it.

# A HUMAN COMPUTER

Chapter Two

In the 1950s, the United States and **Soviet Union** were competing to see who could get to space first. This "space race" couldn't be won without people who knew a lot about STEM. For example, mathematicians were needed to figure out the **calculations** that would make spaceflight possible.

Computers were used to figure out that math, but they weren't the machines we use today. Instead, "human computers" worked to compute, or calculate, the math necessary to win the space race. Katherine was one of those human computers.

## West Area Computers

When Katherine first joined NACA, she was part of the West Area Computing unit. This was a group of Black women hired to work as computers at the Langley Research Center in Virginia. Because of segregation laws, the Black computers worked in the West Area, so they could stay separate from the white computers.

**Fast Fact**

Katherine once said that she worked during a time "when the computer wore a skirt."

Shown here is a photo of Dorothy Vaughn, who played an important part in the beginning of Katherine's career with NACA.

Dorothy Vaughan was the head of the West Area Computers, and she matched computers with parts of NACA that needed their help. She sent Katherine to work for a part of NACA called the Flight Research Division.

## Focused on Flight

Katherine's first years at NACA were spent with the Flight Research Division. Her job was to take information from flight tests and analyze it, or study it to find out more about it.

## DOROTHY AND MARY

Katherine wasn't the only West Area Computer to have her story told in *Hidden Figures*. The movie focuses on Dorothy Vaughn and Mary Jackson too. These women also broke barriers as Black women in STEM.

Dorothy was a supervisor, or manager, at NACA at a time when women didn't often have leadership roles. She was also NACA's first Black supervisor. When electronic computers began to replace human computers, Dorothy became a **computer programmer**.

Like Katherine, Mary worked as one of Dorothy's computers. Then, in 1958, she became the first Black woman to work as an engineer for the U.S. space program.

Mary Jackson (*shown here*) later took a job as the head of the women's program at Langley. This allowed her to support women in STEM by giving them better opportunities in the workforce.

**Colored smoke is used here to show how wake turbulence works. Because of the way the air around a plane is disturbed as it moves, planes have to stay a set distance away from each other. Mathematicians help calculate how far away that distance needs to be.**

**Fast Fact**

In 1956, Katherine's husband died. Three years later, she married a man named Jim Johnson, and they were together until his death in 2019.

As part of the Flight Research Division, Katherine also analyzed a plane crash. It was caused by wake turbulence, which is a **disturbance** in the air behind an airplane as it moves.

## From NACA to NASA

The space race heated up in 1957. That year, the Soviet Union sent the first man-made satellite into space. The United States knew it had to do something big to keep up. In 1958, a collection of notes on space technology written by people at NACA was published, and these notes included math done by Katherine.

That same year, NACA became NASA—the National Aeronautics and Space Administration. It also became a desegregated workplace, although Black women like Katherine still faced **discrimination** at work and outside of it.

## Fast Fact

A satellite is an object that orbits, or travels around, a larger object in space. The Soviet satellite that was launched in 1957 was named *Sputnik 1*.

## An Important First

In the early days of NASA, a group of male engineers were chosen to lead the research on space travel. Katherine often did the math that allowed these men to do their jobs.

In 1960, Katherine worked with an engineer named Ted Skopinski to calculate the orbit, or path, of a spacecraft if the place where it is supposed to land is known. They published their findings, and Katherine was named a coauthor on the report. This was an important first for Katherine, who was about to be a part of many important firsts for NASA.

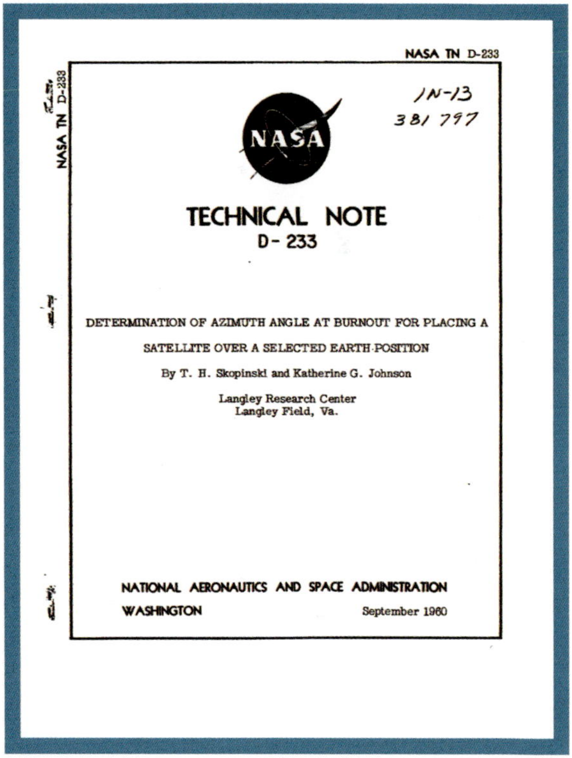

NASA TN D-233

IN-13
381 797

NASA

TECHNICAL NOTE
D-233

DETERMINATION OF AZIMUTH ANGLE AT BURNOUT FOR PLACING A SATELLITE OVER A SELECTED EARTH POSITION

By T. H. Skopinski and Katherine G. Johnson

Langley Research Center
Langley Field, Va.

NATIONAL AERONAUTICS AND SPACE ADMINISTRATION
WASHINGTON September 1960

Shown here is the first report Katherine coauthored, or wrote with someone. She was the first woman from the Flight Research Division to be named the coauthor of a scientific paper.

Space exploration reached new heights in 1969, when American astronauts first walked on the moon. This wouldn't have been possible without Katherine's calculations.

# SPACE STEM

Chapter Three

Sending people into space requires every area of STEM. Scientists need to figure out the **physics** behind space travel and the ways the human body is affected by being in space. Engineers design the technology needed to power spacecraft and keep astronauts alive in space. Mathematicians calculate the trajectories, or paths, a spacecraft can take. This allows astronauts to land safely back on Earth.

Katherine was one of those mathematicians. She helped make spaceflight safe and successful with her calculations.

## Sending Shepard to Space

In April 1961, the first man was sent into space—but it wasn't an American. It was Soviet cosmonaut Yuri Gagarin. NASA knew it needed to send an astronaut to space to keep from falling behind in the space race. The next month, Alan B. Shepard Jr. became the first American in space.

**Fast Fact**

A cosmonaut is the title given to a Soviet (or Russian) astronaut.

Katherine did the math necessary to figure out Alan's trajectory. Her calculations allowed people to find him safely when he landed in the Atlantic Ocean. Because the computers we'd use today for this kind of work weren't around yet, Katherine did all the math by hand.

**Fast Fact**

Alan Shepard's flight lasted 15 minutes. Unlike Yuri Gagarin, he didn't make a full orbit of Earth, so his flight is known as a suborbital one.

## A Successful Orbit

The next year, NASA was ready to send its first astronaut into orbit around Earth. Because this was such a **complex** mission, NASA had set up a network of electronic computers to plot the trajectory of astronaut John Glenn's flight.

However, John didn't trust the machines. He wanted Katherine to

Katherine's math made it possible for people to find Alan's spacecraft, *Freedom 7*, when it splashed down in the ocean, which is shown here.

check their math to make sure it was right. Katherine did the math by hand, and she said the electronic computers were correct. John was good to go on the first orbital flight by an American astronaut.

**Before John Glenn orbited Earth in the *Friendship 7* spacecraft (*shown here*), he wanted to make sure "the girl" checked the calculations. He trusted Katherine more than the machines, even though he didn't know her name.**

## Putting a Man on the Moon

In July 1969, Neil Armstrong became the first person to walk on the moon. This was an incredible achievement for NASA, and Katherine was proud of her role, or part, in making it happen.

Katherine's calculations played a big part in the success of the Apollo 11 mission that first sent astronauts to the moon. She helped figure out the trajectories necessary to make sure the spacecraft that went to the moon's surface could link up with the spacecraft that was orbiting the moon. This allowed the astronauts to leave the moon and get back home.

**Fast Fact**

Much of the math Katherine did for NASA was geometry. This is the branch of math that deals with lines, shapes, and space.

## A Long Career

Katherine worked on many other projects during her long career at NASA. She helped with the calculations that made other moon landings possible. Then, when NASA shifted its focus to the **space shuttle** program, Katherine worked on that too. She even calculated trajectories for a trip to Mars!

Katherine retired, or ended her career, in 1986. However, her work as a STEM **role model** was far from over. In fact, it was just beginning.

Katherine was very proud of the work she did to help astronauts get to the moon.

# MAE JEMISON

Katherine Johnson and many other Black women worked behind the scenes to send astronauts to space. Over time, Black women also stepped into the spotlight as the astronauts being sent on these amazing journeys.

Mae Jemison once said, "Never be limited by other people's limited imaginations." As a Black woman in STEM, she knew that it was important to believe in herself and ignore people who didn't believe in her.

In 1992, Mae Jemison became the first Black woman sent into space by NASA. She was a medical doctor who performed biology experiments in space. These included experiments about motion sickness in space and what being in space does to our bones. Mae is still working today to inspire young people to become interested in STEM subjects.

Katherine is shown here near the end of her career with NASA. For many years, her work—and the work of many Black women like her—went unrecognized. However, people are now beginning to learn about all they did to make space travel possible.

# NO LONGER HIDDEN

Chapter Four

For many years, the American public only knew the names of the astronauts—often white men—who traveled into space. They didn't know the names of the mathematicians—often Black women—who helped those astronauts safely get to and from space.

Over time, though, Katherine's achievements became impossible to ignore. As she was honored for her service to STEM and her country, she became a symbol of all the hard work Black women did—and are still doing—to make fantastic STEM feats happen.

## Honored for Her Work

The Presidential Medal of Freedom is the highest **civilian** honor given out in the United States. In November 2015, Katherine was presented with this medal by U.S. president Barack Obama.

> **Fast Fact**
>
> After Katherine retired from NASA, she spent much of her time speaking to groups of students about the importance of studying STEM subjects.

The next year, it was announced that a new NASA research facility would be named the Katherine G.

## Fast Fact

Having a scientific paper published is a big achievement for someone working in STEM. During her career, Katherine was the coauthor of 26 scientific papers!

Johnson Computational Research Facility. The facility opened in 2017.

When Katherine was asked how she felt about having it named after her, she joked, "You want my honest answer? I think they're crazy." She was proud of the work she did, but she wasn't used to being in the spotlight.

**Katherine is shown here getting the Presidential Medal of Freedom from President Obama. Next to her is baseball legend Willie Mays, who also received the Medal of Freedom that day.**

**Katherine has many things named after her now. In addition to the Computational Research Facility (*shown here*), schools, college buildings, and even spacecraft have her name on them.**

**Fast Fact**

In 2018, Katherine was turned into a doll as part of the Barbie Inspiring Women series of dolls.

## On the Big Screen

Even after receiving the Presidential Medal of Freedom, not many people knew Katherine's story. However, the release of the *Hidden Figures* book and movie in 2016 made her one of NASA's most famous faces.

*Hidden Figures* was a very successful movie. It made around $236 million. It was also nominated for Academy Awards, which are the most important prizes in the movie business. Many people watched *Hidden Figures* and learned about Katherine, Dorothy,

Shown here are the women who starred in *Hidden Figures.* Octavia Spencer (*left*) played Dorothy Vaughn, Taraji P. Henson (*center*) played Katherine Johnson, and Janelle Monáe (*right*) played Mary Jackson.

Mary, and the other West Area Computers. These women were finally getting the recognition they deserved.

## Making a Difference

Katherine lived long enough to see her life's story become an inspiration for many women and girls. She died on February 24, 2020, at the age of 101.

That wasn't the end of her story, though. After her death, her family set up the KCGJ Foundation, also known as the Katherine Johnson Foundation. Its purpose is to provide money for young people to get a good education, especially in STEM subjects.

## Making It Count

Katherine Johnson turned a childhood love of counting into a career as one of NASA's most respected mathematicians. She worked hard, she loved what she did, and she believed those two things were the secret to her success.

## FOLLOWING IN KATHERINE'S FOOTSTEPS

Throughout her life, Katherine worked hard to teach young people, especially Black girls, about the value of a strong STEM education. Katherine loved math, and she believed it was important for all students to have the opportunities to succeed in math the way she had.

Today, many groups are working to support STEM education, especially math education, for Black girls. For example, Black Girls Love Math is an organization that provides after-school programs, math contests, and career days to help Black girls see math skills as a kind of superpower.

Katherine once said, "Like what you do and then you will do your best." Her best helped take us to outer space, and her story reminds us that there are so many hidden STEM superwomen whose achievements are waiting to be discovered.

**President Barack Obama called Katherine "a pioneer who broke the barriers of race and gender, showing generations of young people that everyone can excel in math and science and reach for the stars."**

# TIMELINE

| In Katherine's Life | World Events |
|---|---|
| | **1914-1918** World War I is fought. |
| **1918** Creola Katherine Coleman is born on August 26. | |
| | **1939-1945** World War II is fought. |
| **1953** Katherine begins working for NACA, which would later become NASA. | |
| **1962** Katherine is asked to check the math before John Glenn becomes the first American to orbit Earth. | |
| | **1963** Dr. Martin Luther King Jr. delivers his "I Have a Dream" speech. |
| | **2008** Barack Obama becomes the first Black man to be elected president of the United States. |
| **2016** *Hidden Figures* is released as both a book and a movie, sharing Katherine's story with the world. | |
| **2020** Katherine dies on February 24. | **2020** The COVID-19 pandemic spreads around the world, leading to the deaths of millions of people. |

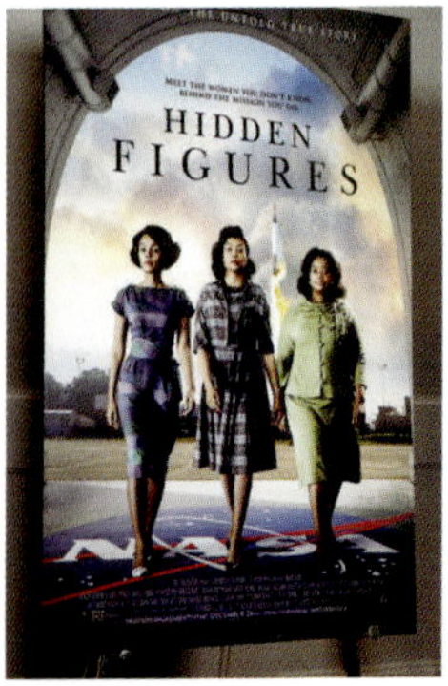

# THINK ABOUT IT!

1. Why do you think the achievements of Katherine and other Black women at NASA went unrecognized for so long?
2. Do you think John Glenn should have used Katherine's name when he asked her to check the math before his flight? Why might it matter?
3. Why do you think many Black women still face barriers to success in STEM?
4. What do you think was Katherine's most impressive achievement?

# GLOSSARY

**calculation:** The process of finding the solution to a problem using math processes.

**civilian:** Dealing with a person not on active duty in the military.

**complex:** Having many parts.

**computer programmer:** A person whose job is to write the code, or set of instructions, that allows a computer to perform certain tasks.

**discrimination:** Unequal—and usually unfair—treatment based on factors such as a person's race, age, religion, or gender.

**disturbance:** A change from what is normal or usual.

**excel:** To do or be better than others.

**physics:** The branch of science that deals with matter and energy, as well as how they interact.

**research mathematician:** A person who closely studies numbers to find new ways to solve problems or analyze data.

**role model:** A person who others look up to.

**Soviet Union:** A country that existed in the 20th century in Eastern Europe and was a major world power before breaking up into many smaller countries, including Russia.

**space shuttle:** A spacecraft that was created to be used many times to transport people between Earth and space.

# FIND OUT MORE

## Books

Hansen, Grace. *Katherine Johnson: NASA Mathematician*. Minneapolis, MN: Pop!, 2023.

MacCarald, Clara. *Katherine Johnson: Barrier-Breaking Mathematician*. Minneapolis, MN: Jump! Inc., 2024.

Orr, Tamra. *Katherine Johnson*. Kennett Square, PA: Purple Toad Publishing, 2020.

## Websites

***National Geographic Kids*: Katherine Johnson**
*kids.nationalgeographic.com/history/article/katherine-johnson*
*National Geographic Kids* presents a short biography of Katherine Johnson.

**Official Website of the Katherine Johnson Foundation**
*www.katherinejohnsonfoundation.org/*
The Katherine Johnson Foundation offers facts about Katherine's life, a timeline of her work, and information about how the foundation is working to keep her legacy alive.

**Who Was Katherine Johnson?**
*www.nasa.gov/learning-resources/for-kids-and-students/who-was-katherine-johnson-grades-5-8/*
Visitors to this NASA website can find the answers to important questions about Katherine's life.

# INDEX